A BOOK
FOR THINKERS!

A BOOK
FOR THINKERS!

TO IMPROVE FAMILY RELATIONS WITH DECISION MAKING METHODS IN A CHANGING WORLD

THE THINKER by August Rodin
Legion of Honor, San Francisco, California

"NEVER STOP ASKING QUESTIONS."
A. EINSTEIN

James Elander

Library of Congress Control Number: 2022922503
ISBN: Hardcover   978-1-6698-5856-0
           Softcover   978-1-6698-5855-3
           eBook   978-1-6698-5854-6

Print information available on the last page.

Rev. date: 03/03/2023

To order additional copies of this book, contact:
Xlibris
844-714-8691
www.Xlibris.com
Orders@Xlibris.com
843970

Objectives

This book tries to help people learn to make better personal and family decisions, via small group discussions, that will lead to improved understandings and better conclusions. The need for conclusions resulted from methods that started in Athens over 2500 years ago in order to aid the first democracy and to improve decision making. It was the first form of a democratic government. We are still trying to improve, but with different tools like computers, TV, etc.

The validity or invalidity of people's conclusions are often based on the information they read or hear, as in advertising, newspapers, TV, and who said it or reported it. Many times, the explanations are incomplete and news reports lead to incomplete or even false conclusions! When all participate, and all discuss possible results or viewpoints, better conclusions by families will result. This book will help all, who read it, to make better future decisions, plus it will improve group or family relationships. To make it even more interesting, bits of related history are added.

Prologue: The Origins of the Ideas

The idea for this book came to the author when he was talking with some men and women who had been looking at and thinking about how to better understand some of today's problems. They said they would appreciate and enjoy activities to help them understand new issues and arrive at better and valid conclusions, but they added to please keep it simple. To create some interesting discussion, I would suggest and urge the father and mother or adults read at least the first 20 pages of Neil Postman's book, TECHNOPOLY.

I said I would keep it simple via a short HOW and WHY book with bits of related history. This makes the topics more interesting and up to date in the current world applying to local and meaningful situations. The activity types are simplified so that all ages can understand, but remember that asking questions is always the key to understanding. Discussions, which are beneficial for not only the adults but for young and old, will help to improve thinking and decision making ability. Getting better understanding is always important to ensure we are making valid decisions. The method will start with simple cases and progress to more complicated problems that you might run into every day. Simple cases or methods will lead you to complicated ones with much better understanding.

We will start by arriving at some conclusions which are from the basic counting number activities. Some will be valid and some invalid.

All of us have encountered the first 10 counting numbers from math, or what are called the basic set of digits. We probably use them more than any other set of numbers. So let us start with that set, the 1-10 whole numbers. A scientific calculator is suggested to help you make some conclusions, plus save a lot of time. In these cases, a general statement is the objective. The idea is for the individual to make a general statement or conclusion from a source which may or may not be valid or true, relative to the general case. The method leading to discussions is the family's or the group's responsibility, then evaluate as to a general truth or interpretation. Suggestion: one night per week called a family night.

Basic Counting

Digital Activity 1: Which integers or counting numbers are divisible by 1? Conclusion: All of this set of numbers are divisible by 1.

Digital Activity 2: Select about 10 integers from 1 to 500 that are divisible by 2 and write a conclusion.

Digital Activity 3: Which of these are divisible by 3: a)12   b)27   c)456?

Now, take each of these numbers, and add up the sum of the digits. You will get:
  a) 1 + 2 = 3
  b) 2 + 7 = 9
  c) 4 + 5 + 6 = 15
For c) above, now add those two digits, and you will get 1 + 5 = 6.

Select a few more below 500, and add them. Look at their sums, then write your conclusions. I suggest you use your calculator.

Digital Activity 4: Do the same as in Digital Activities 2 and 3, but this time use only numbers divisible by 4. Write your conclusions. (Example: 248/4 = 62 which adds to 8 and is divisible by 2.)

Repeat the same **Digital Activities for the numbers divisible by 5, 6, 7, 8, 9, and 10.** Write your conclusions.

Keep your notes, they may be useful later on. Remember, your conclusions are assumptions and may not be valid for the general case. In other words, your conclusion may be true or false, unless you prove it. This book will not do proofs. **There is a difference between conclusions and proofs.**

Note: Mathematics is used to learn many ideas. In this book, answers to problems are given for many cases and some are even wrong. This is on purpose to make you think and prove you are correct!

Suggested selections for SHORT and very INTERESTING READINGS!

FLATLAND, A ROMANCE OF MANY DIMENSIONS
Abbott, E. A.
(This was written about 1890 and is still popular.)

SLICING PIZZA, RACING TURTLES, AND FURTHER ADVENTURES IN APPLIED MATHEMATICS
Banks, R.

HISTORY OF PI
Beckmann, P.
(The question of this constant that started over 2000 years ago and was completed in 1870. Why was the letter π selected and who selected it? You may also be interested in the State of Indiana's case related to pi.)

TAXICAB GEOMETRY
Byrit, D.
MATHEMATICS TEACHER
May 1971

HISTORY OF ELEMENTARY MATHEMATICS
Cajori, F.

BIBLICAL NUMEROLOGY
Davis, J.
(Pi in the BIBLE! See I Kings 7:23)

THE MATHEMATICAL EXPERIENCE
Davis And Hersh

DESCARTES DREAM
Davis And Hersh

MATHEMATICS-THE NEW GOLDEN AGE
Delvin, K.

NUMEROLOGY or WHAT PYTHAGORAS WROUGHT
Dudley, Underwood.

GREAT MOMENTS IN MATH BEFORE 1650
Eves, H.

THE MATHEMATICAL MAGPIE
Fadiman, C.

NATURE OF PROOF
Fawcett, H.
13th Yearbook of the NCTM
(A MUST FOR TEACHERS OF GEOMETRY)

ENGINEERING AND THE LIBERAL ARTS
Florman, S.

MATHEMATICAL CARNIVAL
Gardner, M.

MATHEMATICAL CIRCUS
Vintage Books

NUMBER: FROM AHMES TO CANTOR
Gazale, M.

STATISTICS FOR THE TWENTY-FIRST CENTURY
Gordon, S.

MATHEMATICS FOR THE MILLIONS

THE WONDERFUL WORLD OF MATHEMATICS
Hoben, L.

HOW TO LIE WITH STATISTICS
Huff, D.

HEMHOITZ AND THE NATURE OF GEOMETRIC AXIOMS
Kenny

The WORLD OF MEASUREMENTS
Kien, H.

MITS, WITS, AND LOGIC
Lieber. L.

The EDUCATION OF T.C. MITS
Lieber, L.

WOMEN IN MATHEMATICS: SCALING THE HEIGHTS
Nalon, D. (Editor)

RIDDLES IN MATHEMATICS
Northrop, E.

THE MATHEMATICS OF GAMES AND GAMBLING
Packel, E.

I THINK THEREFORE I LAUGH
Paulos, J.

THE MATHEMATICAL TOURIST
Peterson, I.

THE GOLD BUG
Poe, E.

TECHNOPOLY
Postman, Neil

A LONG WAY FROM EUCLID
Reid, C.

A SHORT HISTORY OF MATHEMATICS
TREASURE ISLAND
Sanford, Vera

TREASURE ISLAND
(Chapter 31)
Stevenson, R.L.

A RANDOM WALK IN SCIENCE
Weber, R.

DONALD DUCK IN MATHMAGIC LAND
(Video or Film)
Walt Disney Productions

CONTENTS

Chapter 1: Reasoning and Conclusions

Activity 1.1: Inductive Reasoning Methods

This method of arriving at a conclusion from a few cases is called **Inductive Reasoning,** which is observing or checking a few cases, and from these cases making a general conclusion. You can naturally see how this is used in many situations. Ask the group to list a few conclusions from a few cases that they have arrived at! An old statement from history is: "Wind from the East is not good for man nor beast."

Another example: A farmer had a pet turkey that always met him each morning for some food. The turkey was REALLY SURPRISED one morning. (Ask the question why?) The turkey made a false conclusion based on past cases. What holiday do you think it was?

Many decisions we make are based on the following:

Inductive reasoning is drawing a conclusion from a few previous cases to the general one. Example: If you are over 21, then you have a driver's license. May be false.

Deductive Valid Reasoning is drawing a conclusion from the general case which is true or has been justified. Example: If you have a driver's license then you are over 21.

a. General Statement: If you have a DRIVERS LICENSE, then you are over 21.
   A valid statement.
b. **Converse:** John is over 21, then John has a driver's license. This conclusion may not be valid. More on this topic later on.

Many times, we may draw a conclusion from what we see. The following is one of my favorite cases dated 1915 in Puck Magazine. What do you see?

Do you see an old woman's face or a young woman's profile?

This is one of most interesting cases illustrating what you see may be false. It was a drawing by W. Hill back in 1915. It illustrates how a different conclusion can be made by observers.

Family Information: Explain to your children or your group how you and your spouse met? Their turn for explanations is next!

Activity 1.2: How do we get information?

Many people get their information from what they read, hear or see. The question is: Are they sure that what they read, hear or see is valid or not one sided? Does what you read or hear cover both sides of the issue with valid statements, and you arrive at logical questions? What you listen to on TV, radio, or lectures, may leave you with interpretations of implications, theorems and laws. The interpretation of converse, inverse, or the contrapositive of statements is very important. This also is used by advertisements and polls. The following diagram is an easy way to assist you to arrive at a valid interpretation of statements.

This picture below is an easy way to illustrate the interpretation of an implication. If A, then B.

There are 4 forms of an **implication**.

Implication is indicated by A→B, which is read as: If A, then B.

There are three other forms of an **implication**, which are given names.

Converse: If in B then in A. (Observe the diagram: If in B, you may not be in A.) See diagram! Converses may be invalid.

Inverse: If not in A, then not in B. (This may not be valid.) See diagram!

Contrapositive: If not in B then not in A. (This is valid!) See Diagram!

The following list indicates the most frequent items that are used and possibly cause people to make false conclusions.

Advertisements or Polls are a very popular methods for getting information to you, but are they complete or valid? Who supplied the information? What was the number involved and how were the questions worded, or are they valid? Are the inductive conclusions always valid? Much of this information is not revealed.

Ask these questions: how was the information collected and what was the quantity, where was it collected and who made the interpretations?

It is suggested that the dinner or post dinner discussions provide an opportunity for the whole family or group to take time to discuss some of these questions, and read the interpretations, plus help expose or question the conclusions. Go ahead and list potential questions, so the whole family or group will understand the problems. (Naturally the question or problem needs to fit the interest and curiosity of the group, a careful selection is needed.)

Activity 1.3: Taxi Routes and Inductive Reasoning

Here is a sample case that can involve even grade school students to help solve and predict the answers.

This Taxi Cab problem is a simple case: The method is to solve a few simple cases and then predict the number of routes for the general case. The cab must always be going in the general direction of A to B.

Ask a person to draw the problem and count the ways or routes the cab can take.

Case 1: Only 2 ways to go from A to B.

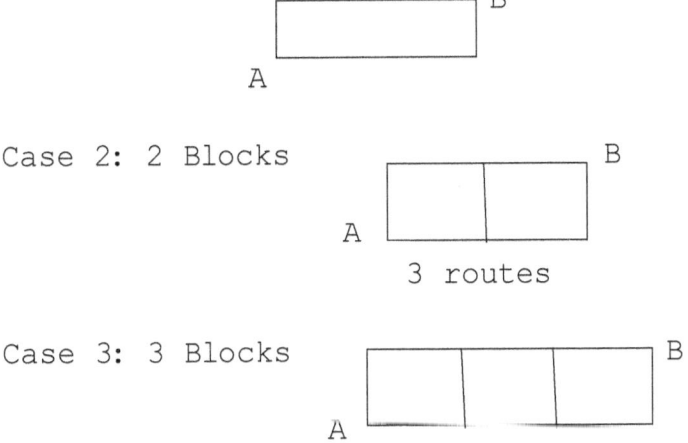

Case 2: 2 Blocks

3 routes

Case 3: 3 Blocks

5

Your guess from cases 1 and 2, how many ways are there from A to B?

Case 4: 4 Blocks
What is your prediction for the number of ways for 4 blocks?

Case 5: 5 Blocks

A

How many ways from A to B? (Predict first using the previous conclusions or answers and then count the ways.) This will check your answers.

Case 6: How many ways from A to B when you have n blocks?

Your predicted answer is _____.
What is the actual count? Comment on your conclusions.

Activity 1.4: PREDICTIONS

The problem is to be able to predict the sum of a set of odd numbers. (Example: What is the sum of the first 5 or 10 odd counting numbers?)

Case 1: What is the sum of the first 2 odd numbers? What is the sum of 1 + 3?

Case 2: What is the sum of 1 + 3 + 5?

Case 3: What is the sum of the first 4 odd counting numbers?

Case 4: Can you predict the sum of the first 5 odd numbers?

Case 5: Write your prediction for the sum of the first 10 odd counting numbers. Comment: Do you see the method?

Activity 1.5: Do you see the arrow?

The following is the picture of the name on the side of one of the company's trucks.
The problem is to see the arrow.
Hint: Look at the X.

FedEx

Activity 1.6: The Mobius Belt Surprise

This an interesting problem and is associated with many actual applications.

Using an 8-inch by 11-inch sheet of paper, cut from the sheet a 1 inch by 11 inch rectangle for each person and call it a belt. Label it as shown.

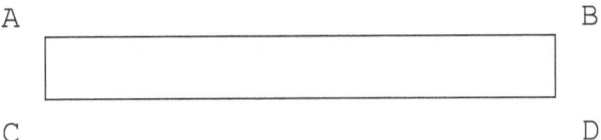

Bend the belt into a circle by putting edge AC against edge BD, but twisting the belt one turn so that A is on D and C is on B. and tape the ends together. You should now have a twisted round belt, with only one twist. Now using a magic marker or a crayon, try to color one side one color, and the other side another color.

Ask: What is the problem? (The belt has only one side!)

Comment: Recognizing the methods in simple problems or activities and then seeing them in everyday cases is what you have to do for practical applications.

Can you give some examples of inductive reasoning for discussions?

Here is one example: The school bus always arrives each school day at 7:30 AM. This Tuesday it didn't! The statement was false!

Another example: All patriots fly the flag on Dec. 7th. Mr. G does not fly the flag on Dec.7th, therefore he is not a patriot…or can he still be a patriot?

Discussion: What do you recall that you thought at one time was true and is not true?

Activity 1.7: Point-Line-Region and a surprise.

Give each person a sheet of paper and ask each to draw 6 circles on their sheet with about an inch radius. In the first circle indicate 2 points on the circle. Question: You may ask what the definition of a circle is?

(A circle is the set of points on a plane equal distance from a given point on the same plane.)

Conclusion: 2 points, A and B, on a circle and the line segment AB divides the circle into 2 regions or areas.

2 Points make 2 regions.

3 Points on the circle and how many regions? On the next circle, make 4 points and how many regions do you predict?

Complete cases for 4, 5, and 6 points.

What are your predictions for the regions for each number or points?

In the first sets of activities, you and the family or group have been shown that some conclusions may be true, but you are not sure for the future cases. The next case may be false.

Comment: Can you think of any other cases?

Activity 1.8: Course information

If you are a student in high school, the following problem may be of special interest to you. The difficult question may be to answer what you might want your major to be in college. The answer may change several times in the 4-year period. Hopefully, there is no question as to the **need** for an advanced educational program.

The following program will help you to select or consider a possible major. Select a number from the following list that you think could be a major you might enjoy.

1. English
2. World History
3. Biology
4. Physics
5. U.S. History
6. Chemistry
7. Athletics
8. Social Work
9. Mathematics

After your selection, perform the following steps which will help you to be best qualified for your future.

1. Select the number of the potential major and multiply it by 3.
2. Add 3 to your answer in number #1.
3. Multiply the answer to #2 by 9.
4. Add the digits in the answer in step three until you get a one-digit number.
5. Match your answer in step 4 with the above listed subject/major list. This will tell you what the most beneficial subject is for your chosen major.

This is an example of a rigged response. The answer is always mathematics for the above example. I thought you might enjoy it, but on the other hand, many polls and advertisements use this rigged response technique to get you to come to a false conclusion.

Write a summary of chapter 1.

Discuss with your group or family some memories of early years. (If you have some early pictures of your family and relatives, where you lived, your wedding, etc., share them with the group. This could be very interesting for the family or group.)

What comments or conclusions do you have?

Chapter 2: Decisions

Activity 2.1: What Decisions are Based On!

You may find it convenient to discuss how your family, or even persons in your group, individually make decisions. How do people make decisions and what are the decisions based on? The responses may be very interesting.

All decisions are based on **undefined terms, definitions, postulates** or **assumptions,** and **previously agreed to conditions, like laws.**

About 40-50% of what we say, write or read consists of undefined terms. Many people question the 40%-50% and the words that are undefined. A great statement, often used to illustrate this, is the preamble to the United States Constitution.

"We the people of the United States, in order to form a more perfect union, establish justice, ensure domestic tranquility, provide for the common defense, promote the general Welfare and secure the blessings of liberty to ourselves and our posterity, do ordain and establish this Constitution for the United States of America."

Have someone in your group read the Constitution's preamble aloud, and then the group can classify each word as definable or undefinable. (How many did they classify as undefined?)

What is the percent of the total number of words in the preamble that are undefinable?

N% = (Number of Undefined Words divided by the total number of words) times 100 and add percent symbol (%).

Example: Assume there are 10 words classified as undefinable and 20 words as definable, therefore (10/30) times 100 and add the % symbol%. Which now reads as 30 % of the words are undefinable.

The dictionary tries to do an impossible job! A word is definable if the definition is valid when reversed.

EXAMPLE: A DOG is an animal with 4 legs. True, but not a definition. Reverse it and it is false.

There will be much more on this topic as you progress with new material.

Many decisions are based on the following:

    GUESSING
    ILLUSIONS
    QUOTES from experts or partial experts
    READING books, fiction and non-fiction
    INDUCTION, direct or indirect
    ESTIMATING
    STATISTICS
    IMPLICATIONS (including Converse,
            Inverse, and Contrapositive)
    MEDICAL REPORTS

CONTRACTS (such as insurance)
ADVERTISEMENTS
LECTURES
RADIO (including Talk Shows)
TELEVISION
EDITORIALS

The text now will have questions and problems to be discussed, explained, solved and hopefully create discussions which will improve critical decision-making skills.

List your comments and questions!

Activity 2.2: Banker's Rule for counting money.

It was the Chinese bankers about 1000 BCE that informed the European bankers that they had to use the "MDAS" method for evaluation of amounts of money. The MDAS is translated as **multiply** then **divide** followed by **add** then **subtract** or basically multiply then add, when evaluating an amount of money. This is a method of counting money that the ancient bankers of EGYPT (1000 BCE) demanded be used in a certain way. A simple case using money is: $8 + 3(6-2)$. One way to do it will lead to: $+ 8 +18 -6$. What is your answer? (The correct answer is **20.)**

This evaluation order that the banks demanded is MDAS, multiply, divide, add and subtract or really multiply first and then add since division is a form of multiplication and subtraction is really adding the negative. (Many teachers in some of the early grades translated this as "**M**y **D**ear **A**unt **S**ally".)

The easy way to understand the why for this is to evaluate the coins in a piggy bank, say you have 4 half dollars, 10 quarters, 13 dimes, 15 nickels and 40 pennies. How much money is in the piggy bank?

For more history about the Chinese influence on banking in Europe, read the book **1421** by Gavin Menges. The most important function of a leader is to, in a unique way, motivate others to **question** and **understand** problems.

Question: Why can't we divide by zero?
N/0 =? What are some of the possible answers and then try to check your answer.
Your comments:

Topics for family discussion: Insurance -- House, medical, car, life. Invite local agents for each type to answer questions.

Activity 2.3: How to become a millionaire in 30 days.

Ask your group or friends if they would like to be a millionaire in 30 days? Their answer may be interesting as to what and how, why or why not! If they say yes, they will probably add, "HOW?". Tell them you will do the deposits the first 20 days and your friend the last 10 days. The method is simple: You will deposit 1 cent the first day, 2 cents the second day, 4 cents the third day, 8 cents the fourth day (as you can see, the deposit amount doubles each day), and so forth for 20 days and your friend the last 10 days. (A good deal for your friend?)

Write out the procedure:
Day: 1 2 3 4 5 … and so to 20.
Deposit: 1+ 2 + 4 + 8 + 16 + . . . to the 20th.
Total sum: 1, 3, 7, 15, 31, . . . total value.
The total is $1,342,177.28
Your comments or your friend's comments?

Activity 2.4: Easy Sums

What is the sum of the first 10 odd counting numbers? Hint: Try several simple cases and look for a pattern.
Case 1: 1 + 3 is 4
Case 2: 1 + 3 + 5 is 9
Case 3: 1 + 3 + 5 + 7 is 16
Can you now predict the sum for the first 10 odd numbers?

Comment:

Activity 2.5: Rule For Doubling Money

This rule is referring to Compound Interest. Use your calculator! Do you understand what Compound Interest is?

Case 1. Calculate (using your calculator) how long it takes money to double at the rate of 8%.   Answer: 9 years

Case 2. Calculate how long it takes money to double at 9%.   Answer: 8 years

Where does the 72 come in?

Notice 72/8= 9, 72/9 = 8, and generalizing 72/r = years to double.
Write the rule.

Comment:

Activity 2.6: Basic types of Thinking

All decisions are based on four classifications which are:
1.   undefined terms
2.   defined terms
3.   basic assumptions
4.   previous conclusions.

The basic model is the Preamble to the U.S. Constitution. The following will help you understand this.

"We the people of the United States, in Order to form a more perfect union, establish Justice, ensure domestic tranquility, provide for the common defense, promote the general Welfare and secure the Blessings of Liberty to ourselves and our Posterity, do ordain and establish this Constitution for the United States of America."

Are there any words in the above that need defining? Pick out a few. Example: common defense.

Select an article from the local newspaper and find a few words that need to be defined in order to really interpret the meaning of the article.

Activity 2.7: The Methods

There are several ways to arrive at a conclusion. The two logical methods are labeled the **direct method** and the **indirect method**.

The direct method is used when the conclusion follows directly from a formal decision.

Example: Direct Method: Joe is speeding and is given a ticket. Conclusion: If you exceed the speed limit, then you get a ticket.

Example: Indirect Method: Say there are 5 possible cases and only one is correct. If 4 cases are eliminated, therefore the last case is the correct one.

Use the **Indirect Method** for the following.
One method to solve thinking problems is by the indirect method, which is to consider all possible cases and if they all prove impossible but 1, then that last answer must be the correct one.

Example: You misplaced your keys so you go back to all the places you have been this morning, and the keys were in none of the locations except where the telephone is, and sure enough, the keys were there.

The following is a more complicated case but the concept is the same. Pete, Joe and Tom were accused of causing a problem at a school party. The principal asked the math teacher to help solve the problem. Each student agreed that of the three statements, only one is true.

Their statements are:

Pete said: Joe did not do it.
Joe said: Tom did it.
Tom said: Joe is lying.

Who did it?

Write each of the responses when one student's statement is assumed true and the others are false statements, then the guilty one will be revealed.

Example or method:
Assume: Pete's statement is true.
Then you have (True statement is in **Black**):
Pete: **Joe did not do it.**   T
Joe said: Tom did it.   F
Tom said: Joe is lying.   F

Assume: Joe's statement is true.
Pete: Joe did not do it.   F
Joe: **Tom did it.**   T
Tom: Joe is lying.   F

Assume: Tom's statement is true.
Pete: Joe did not do it.   F
Joe: Tom did it.   F

Tom: **Joe is lying.**   **T**
Your Comment:
Answer: Joe did it.

Family Discussion: What do you remember as the earliest event in your life and your children's lives. What do they remember?

Chapter 3: Thinking Activities

Activity 3.1: Letter to parents

Henry is a student at the state college and as it usually happens the student needs more money than expected. So, he sent the following email request.

```
   SEND
 + MORE
   MONEY
```
Thanks, Henry

The father and mother think a while and decide the student sent the request for a certain amount and probably each letter is a hidden single digit number. So, all they will have to do is replace the letters with digits.

Hint: Each letter can only be used by one digit. So, if there are two Ns and if one N is 5 then the other N is also 5, and no other letter can be 5. What digit must M be and why? What is the final amount the student is asking for?

Comment:

Activity 3.2: ROMAN NUMERAL for 10

What are the Roman symbols for the numbers 5 and 10?

Using your left hand closed, indicate and look at the following:
   a. Indicate the number 1. One FINGER

b. Indicate the number 2. Two FINGERS
c. Indicate the number 3. Three FINGERS
d. Indicate the number 4. Four FINGERS
e. Now indicate the number 5. What do you notice for 5?

Do you see the "V" which became 5, and 2 V's (fives) make 10, therefore X is the symbol for 10. (See the 2 V's in x, hence x is 10.)

Family discussions: The history of your state or county. (Invite a history teacher!)

Comments:

Chaper 4: Money And Mathematics

Activity 4.1: Simple Interest

This type of interest was used by people who borrowed money from a person before there were many banks, usually a paper contract between two individuals which states the names, dates, interest rate and length of time for a loan.

Example: Joe Friend, borrows from Jack Moneyman the amount of $300 to be paid back on September 1, 1845 (it is now 6/1/1845) at the interest rate of 6%. What is the value of the debt plus pay back?
$= 300 + 300(.06) = 300 + 18 = $318
(For better understanding create more additional cases.)

This is an example of a typical money problem using simple interest before banks.

Activity 4.2: Compound Interest

With banks and computers and calculators, compound interest problems are much easier, but the method of paying interest on deposits is much different and more complicated. It is probably less understood by many people.

Example: John Friend deposits $4000 at 5% and would like to know what the amount will be equal to in 2 years.
Answer: The value is $4000(1.05)^2 = 4410
(Any questions?)

Compound interest means interest is paid on the loan and interest each period. Unless the contract reads differently, the rate is at a yearly rate. Other times, contracts compound daily or monthly. Perhaps you could try compounding yearly, monthly, or daily, and see if it makes a difference! Calculate the final value if the rate is 9% compounded per year and the original deposit is $2,000. Assume the deposit was invested for 8 years. Answer: $3985.13

Compound Interest Formula is:
A = a*(1+r)^n where a is the $ invested ($2000)
    R is the interest rate per year
    N is the number of years
    A is the final amount ($3985.13)

What type of compounding do you like (annually, semi-annually, monthly, or daily) in order to maximize your earnings?

Activity 4.3: Banker's Rule of 72

How investments or debts can be estimated, some history…

This rule probably dates back to the 1800s when banks were becoming more useful than under the bed for your money. Here are two examples for your discovery of the rule of 72. Use your calculator.
1. How many years does it take $100 to double at 8% interest compounded annually?
2. How many years does it take $500 to double at 9% interest compounded annually?

3. Write your conclusion for the general case from 1 and 2 above. From these 2 cases, the answer is the number of years for money to double is 72/(rate of interest.)

A very interesting and informative book for the history of interest is **1421** by Gavin Menges.

Family discussion: Checking accounts, savings accounts, insurance and annuities, life and medical, and invite a financial agent. Another night for family information, since most young people are interested in this.

Activity 4.4: A few everyday problems

How can the following be?

Statement: 1 yard = 36 inches. Divide each side of the statement by 4. What do you have?

Now, take the square root of each side of the statement, therefore: 1/2 yard is 3 inches.

WE KNOW THIS IS NOT CORRECT, BUT WHY?
Your comments? (Hint: this is a true statement yet not an equation.)

Why do some 4-Legged chairs WOBBLE?

Many students feel that their math course did not have enough of the real applications and they are probably correct. Here is an application that the grandparents will probably understand also. Take a flat object like a plate, a book, or a piece of cardboard. Can you support it with one finger?

2 finger tips?
3 fingers? (Or two fingers and a thumb of the same hand.)

Show the cases for the above and explain your conclusion.

Explain why an older person many times uses a cane.

Why may a 4-legged table wobble?
Four points could determine 4 planes.
ABC, ABD, BCD, ACD, where the letters are points (like the thumb and 2 finger tips).

How many planes can 5 points determine?

Activity 4.5: Line Segments from N-Points

This is a similar problem to the point-plane problem, but this time it is points and line segments.

Euclid's postulate states that 2 points will determine one line, so is there a relationship as to lines and points? Why did Euclid write the first geometry book and when? You will be surprised!

Points	Lines
2	1
3	3
4	6

```
5          ?
6          ?
```

Is there a relationship for prediction as to n points and the number of lines?

n Lines = $n(n-1)/2$ = ?

Your comments:

Activity 4.6: Review of Chapters 1, 2, 3

You may want to review a few of these, as after dinner topics. It is advisable to review each chapter with discussions and actually working some of the problems again as needed.

Math Induction: Direct and **Indirect** methods for arriving at a conclusion with some weaknesses pointed out. To get the method across and the understanding that all conclusions are based on undefined terms, definitions, assumptions, and the resulting conclusions. The example used many times is preamble of the United State Constitution.

Many conclusions are arrived at from the use of deductive or inductive reasoning. Which one is usually always valid? (Deductive, why). Which one is often false? (Inductive, why). These interpretations and applications will be emphasized throughout this whole book. A few of the following may be helpful.

Comments:

One of the hardest concepts to understand is **UNDEFINED TERMS**.

One of the great contributions of the United States is the Constitution. After the meal, ask each person to classify each word in the Preamble as definable or undefinable. Then ask each person why the selected words are classified as undefinable? The dictionary tries to do an impossible job. This illustrates why what we read or hear does not provide a clear statement.

Retirement Thoughts: Listen to all!
When, where, why, and what will you hope you do, or would like to do in the future?

Activity 4.7: Happy Days

Many people are encouraged to invest in annuities or other programs as a logical way to prepare for the dream of retirement days. It is wise to dream and prepare for them! Let's look at a simple case.

Assume when you are 35 and you and your spouse take out an annuity and plan to begin to use it at age 65, or when you retire. The plan: You and your spouse pay in $1200 per year for 30 years. Your thought now is: What will this investment amount be when one of us is 65? The interest rate is now 12% by the investment company. This is a good rate since the company usually has the money invested at 15% so just 3% extra kept for the investment company. The first $1200 in 30 years will be **$35,951.91,** and the last $1200 will be **$1,344.** (Do you see where these amounts came from?) Therefore: L = $1,344 and A = $35,951.91. The average (first payment plus last payment divided by 2) will be then **$1,244.34** and the value invested plus interest per year for 30 years will be:

A + L = $35,951.91 + $1,244.34 = $37,196.25
(A + L)/2 = $18,598.125 times 30 = $557,943.75

Hence: At age 65 the total is $557,943.75 for retirement happy days!

Discussions, such as potential future plans from each person: It may be also a good time to look at each of the activities in chapter 2 to re-enforce their meanings and answer any questions like, what is the difference between direct and indirect reasoning? Other comments or questions?

Chapter 5: Interesting Problems

Activity 5.1: Curves from Straight lines

This is an activity that even the very young students can take part in. It is very important to involve the young people and encourage them to ask questions and take part in the activities. This question is: How can you draw a curve using only straight lines?

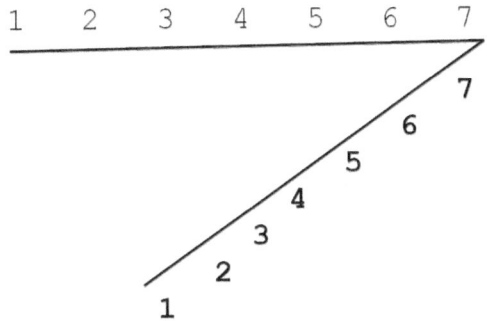

Now draw segments 1 to **7**, 2 to **6**, 3 to **5**, and so forth. By changing the angle of the 2 segments, different curves, even circles, can be formed.

Comments: Try it with a square.

Activity 5.2: Paper Folding

This could be an interesting thinking problem. Ask your group to think about this problem before they give their answer.

The question is: Given a rectangular piece of paper, even a page from the newspaper, how many times can you fold it in half?

No folds   One fold 2 folds

[ ] [ ] [ ]

Record your prediction and then actually count your number of folds. Try it even with a newspaper size page. You will be surprised.

Your comments:

Activity 5.3: The Center of Gravity

The last few problems have been the family type, or more for the younger groups, but still informative for older groups. You may find the next also informative for all, especially the younger group.

The leader, in preparation, should make three figures out of card board: a square, a rectangle, and an equilateral triangle. Make a set for each person, also one set for the leader. Using the tip of a finger (or a pin) try to find the balance point for each of the above. Mark that point, calling it the center of gravity. Write down your conclusion.

   a. Now draw the diagonals in the square and the rectangle and write a conclusion.
   b. Draw the three altitudes for the triangle and write the conclusion related to the center of gravity.
   c. Hold the triangle vertical to the table plane and rotate it clockwise carefully. What do you notice when the point is just

beyond the vertical line to the plane at the vertex indicated. What do you conclude? If it were a car, what would you conclude? Write your conclusion.

Center of Gravity

Tipping line

Where do you think the tipping point is in a car? When the center of gravity is beyond the tipping point line in the car, the car will tip or roll over. If you have a toy car, demonstrate for your friends.

Comments:

Activity 5.4: How to Improve?

Improve your digital writing?
   a. Write the digits 1 to 9, and circle the one you write the poorest.
   b. Multiply the circled digit by 9.
   c. Multiple the answer to b by 12345679.

Are you surprised?

Comment:

A Unique Number?
   a. Write the numbers that are the factors of 28.

b. Add the factors in "a". This number 28 is classified as a perfect number. Why?

c. If 28 is the second perfect number, then what is the first perfect number?

d. Research: How many perfect numbers have been discovered? Guess first and then check the internet.

Your comments:

Activity 5.5: Ways of writing the first 10 digits.

Given four 4s, write them to express the first 10 counting numbers.

a. $1 = 44/44$ b. $2 = 4/4 + 4/4$

c.   d.

e.   f.

g.   h.

i.   j.

Your comments:

Summer Employment
There were three students and, fortunately, each had a summer job. For the first month, Pete earned $500, Don earned $1500 and George earned as much as Don and Pete together. The last paycheck was for the final two months, but the boss cut only one check for the total amount that needed to go to all three. Assuming their rate of pay was the same all summer, and using the information above, how much should each person receive from the final check?

Don gets $?
Pete gets $?
George gets $?
What was the value for the final check?

Activity 5.6: A SUMMARY

Up to this point, the major function of this book has been to refresh your memory about logic and thinking problems you may not have thought about in a long time, with an emphasis on methods to aid in decision making. Remember, all decisions are based on four items: undefined terms, defined terms, assumptions, and conclusions that were based on the first three. You were also exposed to some properties of numbers, a few monetary problems, the differences between simple and compound interest, a simple way to calculate payments over a length of time, and making decisions from cases. Remember, Geometry has been taught over the years (3000) as a decision-making study for better leadership, and to improve the quality of life.

A major objective of this book is to help all to become comfortable asking questions, especially young people, and for older people to share their knowledge they have gained over the years.

One way to review at this point is to re-work the problems that were troublesome in the previous activities. Another way is to work related problems suggested by members of the group that might be more difficult than those above, or be taken from their real-life experiences.

Look at and discuss pictures related to family and relatives with comments.

Student Problem: Preparation for a job

Pam, a senior at the university, is home for the summer and was employed by a local company for 8 weeks in the area that she is majoring in. Her pay for the summer would be $1200 per week. She planned to save $1000 each of the 8 weeks, and at the end of the summer invest the $8000 for her final year and employment hunt.

At the end of 8 weeks, she has $8000 in her checking account. Now she transfers the $8000 into a savings account at 9% interest. She plans to take it out in one year, when she begins her new position. What will the value of $8000 be equal at the end of one year? (Answers: $8720) The compound interest formula is on page 33, if needed.

Question: What would it be in 16 years if she had put the $8000 in a program that paid 9% for 16 years? Ending balance?

The quick answer is using the Bankers Rule of 72, which is divide 72 by the rate of interest and the answer is the time it takes the amount to double. (72/9 equals 8 which means it will double in the first 8 years.)

Answer: Year 8 the Account will be about $16,000. What would the amount be after the second 8 years?
(Check your answer using the compound interest formula. See Activity 4.2.)

Family discussion: Invite a broker for after dinner family discussions and questions relative to debts, investments and other investment plans.

Comments and questions:

Chapter 6: Implications: Their Uses and Interpretations

Many people misinterpret implications and its four forms (**original implication**, **converse**, **inverse**, and the **contrapositive**.) They misinterpret the truth or falsity of the statements and/or the validity of them.

Example: The statement is: If A then B. Also written as A implies B. This is valid.

The **converse** is: If B, then A (B implies A). This could be False.

The **inverse is**: If not A, then not B (not A implies not B). This could be false.

The **contrapositive** is valid: If not B, then not A (not B implies not A). This is valid.

The explanation above is easier to understand by the following diagram for

    A → B:

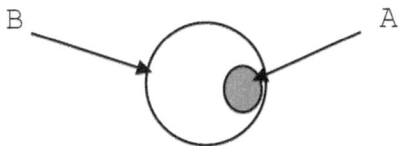

Original statement: If A, then B is true. Do you see A implies B in the above illustration? It is easier to interpret where B is the large circle and A is the small circle.

Converse: If B→A, means if in B, then in A, which could be false.

Inverse: If not in A, then not in B. This could be false.

Contrapositive: If not in B, then not in A. This is true and is valid.

Remember that the original statement, if in A then in B is True and Valid therefore the contrapositive is true and valid.

Discussion: Bring in some ads and state the intention of the ad in if-then forms and interpret the four forms. What are the four forms? Many people misuse or misinterpret statements, especially the converse and inverse.

WRITE YOUR SUMMARY with some examples!

Chapter 7: REVIEW for CONCLUSIONS

Activity 7.1: Banker's rule of 72

Examples: $5^2 = 25$
$(15)^2 = 225$
$(25)^2 = ?$

Do you see the possible pattern or easy way to arrive at the answer? The following will help you see the method.
$(35)^2 = ?$
$(45)^2 = ?$
$(55)^2 = ?$
and continue using the above method
until you see the easy way, but you can't be sure the future cases are valid since you have not proved the general conclusion.

What is the Bankers Rule of 72? An investment of $1000 at 6% will be double the value in 12 years. This rule gives only an approximate answer. Let the group select a few more rates like 8, 9, and 10 percent and determine the results, then write your conclusion. Use your calculator.

What is your conclusion?

Activity 7.2: The Sum of EVEN Numbers

Here is another problem for you to look for a pattern, so that you can predict the answer in other cases or even the general case. What is the sum of n even numbers? $2 + 4 = 6$

```
2 + 4 + 6 = 12
2 + 4 + 6 + 8 = 20
2 + 4 + 6 + 8 + 10 = ?
```

Comment: Do you see the easy method for the answer for the sum of the first n even numbers? If not continue a few more cases.

Discussion: When was the last time you visited a State or National Park, or a local historic site? What do you recall?

Discussion: Give each person time to recollect. Write summary.

Comments and questions:

Chapter 8: Thinking Problems

Activity 8.1: Sum of the Angles

Let's assume everybody knows the sum of the angles in a triangle is 180 degrees. See the first figure below which shows a rectangle has 4 right angles and can consist of 2 triangles, and therefore a triangle has 180 degrees.

Then, what is the sum of the angles in these figures? Hint: Form triangles.

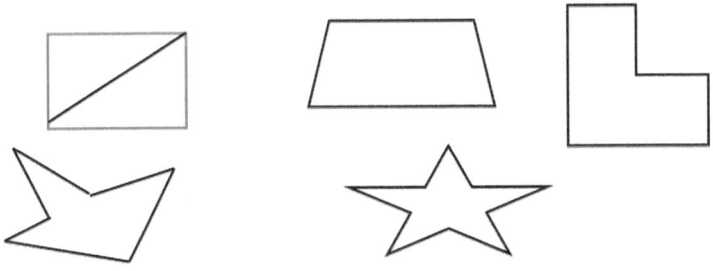

What is the sum of the angles in an n-gon? (An n-gon is a polygon with n-sides.)

Comment:

Activity 8.2: Arrangements

The way 2 pictures can be arranged!
A concept all should understand! We will start with hanging pictures. A person wants to hang pictures horizontally on the living room wall above the sofa. How many ways can she hang them? The answer is 2, AB or BA where A is one picture and B the other picture.

What if she had 3 pictures? A,B,C.
Ways to hang: ABC, ACB, BCA, BAC, CAB, CBA.

How many ways for 4 pictures?

Look for the pattern or the total ways to arrange them.

Pictures	Ways
2	2
3	6
4	?
5	?
N	?

Predict first, then count the number of ways.

Write your conclusion for n pictures!
If you have n pictures, how many ways do you think they could be hung, horizontally?

Comment: The math symbol for multiplying the numbers from 1 to 4 is 4! or 1x2x3x4 = 24. What does N! mean? Now rewrite the answers for the above using the ! symbol.

Activity 8.3: The Question of Probability

You hear and read a lot about the probability of something happening, but we really need a definition of the term, probability. Ask your family or others what they think the definition is.

Definition: The probability of n is n/T where n is the number of favorable results divided by T, the total number of possible results.

Example: Throw a cubic dice with the sides numbered 1 to 6 and you want a 1 to turn up. From the definition in this case the probability of a 1 is 1/6.

1. What is the probability of throwing not a 1? P (not 1) is 5/6.
2. What is the probability of drawing an ace from a deck of cards?

Ask each person in your group to make up and ask each other a related probability question!

What do you think a fair bet is?
Think a while! Then explain that a fair bet is when a win pays what the total cost to play is for all possible results.

Simple case for a fair bet: the dice problem. Say you toss a die 6 times. In theory a 3 should turn up once. Therefore, if one play costs $1 and a 3 should turn up once in 6 tosses, hence 6 tosses will cost $6 and the win should turn up once. Hence a win should pay $6.

Does your state have a lottery and if so, what is the probability of winning? Is playing the lottery a fair bet?

Activity 8.4: Combination Lock Problem

Another counting problem which many students encounter is: how many ways can a 3 number combination be entered to open a combination lock on your locker door?

Assume there are 10 numbers on a dial lock. (Let's assume no numbers are repeated in the combination.)

The first number: 10 choices.
Number 2: 9 choices
Number 3: 8 choices
Probability for selecting the correct 3 numbers is 1/10 x 1/9 x 1/8 = ?

Here is an interesting rule that makes solving probability problems much easier. The rule is: If the probabilities are connected by **"or"** then add them, if by **"and"** you multiply them.

  a. Assume you have 2 throws with a pair of dice, and a win is when both 4 and 6 turn up. What is the probability of a 4 and a 6 turning up? In this case they are connected by **and** so we multiply! Probability of 4 **and** 6 is 1/6 times 1/5 or 1/30.
  b. What is the probability of a 4 **or** a 6 turning up? (1/6 plus 1/5 is 11/30.)

Let each person make up a problem tossing coins and their probabilities, and then actually toss the coins.

Comment: A casino director told me a few years ago that their machines were programmed to pay out about 10% of the take. What do you think about this?

Activity 8.5: Sides of a triangle and their measure, plus are all circles the same size?

1. What do you know about the length of the unknown side of a triangle if 2 sides are given? Given: $a = 7$, $b = 10$.

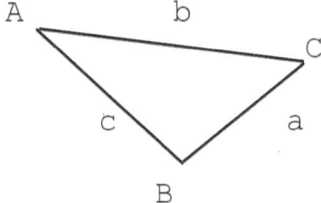

2. In a triangle, can you explain the following logic statement: if side a + side b is larger than side c, then a > c-b.
3. In a triangle, can you explain this statement: If b + c is larger than a, then b > a-c.
4. In a triangle, can you explain this statement: If a + c is > b, then a > b-c.
5. Now if you know a is 7 and b is 5, in a triangle, then what do you know about _____ < c < ____?
6. You know that in a square, the sum of the angles is 360 degrees, but what is the sum of the angles in the following figure?

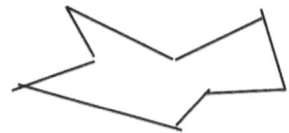

7. Which is larger? 2 or 6?
8. What is half of 8? 0 or 3 or 4?
9. Two different size circles have the same center and are pasted together.

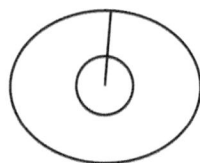

Therefore, when the large circle makes one revolution the smaller one does also. Hence, they must have equal circumferences. How can this be? How can you solve this question? (Actually, make the figure and trace the two points as the large circle makes one complete circle, you will be surprised!)

Activity 8.6: The Question of Volume

The question of volumes for solids and how volumes are calculated is a very interesting one. We will now look at the number of cubic units it contains and how to calculate it.

A cubic unit is like an ice cube. (1 unit of volume is like a box or cube with one unit of measure on each side.)

V = LWH = 1 cu. unit
and all are in the same units.

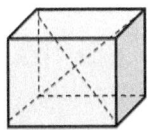

Do you see that a volume of a cube consists of 6 pyramids? Therefore, each pyramid volume is 1/6 of the volume of the cube and the altitude of each is ½ the measure of the altitude of the cube.

The volume for figures like cones, cylinders and spheres is more difficult.

Volume of a pyramid is (1/6)LWH or area of the base times height, where Bh is the volume of the cube, but the height of each pyramid is (1/2) of H or h. Let's call it **small h**, since a pyramid is ½ of the height of the cube. Therefore, the volume of one pyramid = Bh/6, or 1/6 of the volume of the cube, or B2h/6 or Bh/3. Hence the formula for the volume of a pyramid is V = (1/3) Bh cubic units. The formula is the same for a cone but the base is a circle, so the volume formula for a cone is:

$$V = (1/3)\pi r^2 h \text{ cubic units}$$

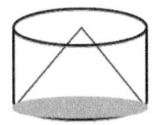

The base area is π times r² and the height is 2r with a cone volume 1/3 of a cylinder with the same base. This means the Volume of a cone is V= (1/3)Base times the altitude (h) or V=(1/3)

$\pi r^2 h$. (The base is a **circle**, so the area of the base is πr^2.)

(This is justified in my 2022 PLANE AND SOLID GEOMETRY, ESENTIALS text, which is now available for schools.)

This activity has another objective and that is to aid a person to draw three dimensional figures on a sheet of paper and insert the dotted lines to indicate the hidden lines. My computer could not do it! All should draw the following: a sphere, box, rectangular solid, cone, pyramid, and cylinder. This could take several days for younger students.

Write your Chapter Summary.

Possible information and questions for family discussion:
- How did your city get its name and when?
- What is the history of your church?
- What is the history related to the local schools and their names.
- The history of the local library.
- Invite a local historian for an evening discussion concerning history.

Keep your notes!

Chapter 9: Family Information

This is a very important topic for the family, especially as the young family members mature and begin to ask questions. The answers which will be passed on are most important! It may be advisable to take only one topic per week for discussion.

Draw a family tree (with pictures, if available) and try to have answers as to where, when and from. You may want to investigate having Zoom® meetings, if some of your family lives far away, and you want to include them.

Activity 9.1: Insurance

Life insurance, medical insurance, car insurance, house insurance, cash values, etc. are great topics to discuss with your family members. This could take several weeks if one topic per meeting is discussed. Invite local agents to answer questions and give more details, such as cash values and yearly reports and needs.

Other topics such as savings, income, investments, charities, church, tax (State and Federal) could also be shared and discussed.

A very special and important topic is family history with pictures and possibly an annual get together or Internet video meetings.

Another great topic would be local, state and National Parks. Suggestion: One topic per week and there should be much discussion with pictures. This could take several weeks, but is worth it.

Comments:

Activity 9.2: Family Topics and News

You can pick up any newspaper and you will find many articles, statements or ads which will require definitions and further information.

Examples:
1.   There is a need for affordable housing. Which word needs defining?

2.   A nationwide poll claimed 80% of those polled want gun control! But only 800 people were polled. It did not say where, when or how, nor any info as to their background.

3.   Drink a lot of water each day. What word needs defining?

4.   People don't walk enough each day. What words need defining?

5.   What are the 4 forms of an implication.

   a. A -> B
   b. the Converse is?
   c. the Inverse is?
   d. and Contrapositive is?

Your comments:

Activity 9.3: Review

1. What is the numerical evaluation for the Order of Operations?
   a. 7(2) + 10/2 − 15(1/3) −6(4)= ?
   b. MDAS means what?

2. In triangle ABC where side **a** is 15 units, side **b** is 10 units, what can you say about side **c**? Complete the following:? > C > ?

3. If you invest $15000 at 8% per year then what will it amount to in 9 years? Answer: about $30,000 or $32578.40 (compounded annually).

4. Discuss your interpretation of selected ads from a magazine or newspaper. (It could be very useful and more meaningful if the leader or each person selects an ad in advance.)

5. All conclusions are based on 4 items! What are the 4?

6. What is the sum of the first 50 odd integers?

7. What is the probability of tossing a head with a coin? (One toss)

8. Probabilities:

   a. What is the probability of tossing a total of 10 with two dice? (1 toss)

b. What is the probability of tossing a head and not a head in two tosses with 1 coin?

9. What is the sum of the internal angles in this figure?

10. If the cost of a 5 inch circular pizza is $5, then what should the price of a 10 inch circular pizza be? (Based on area.) Answer: ($20)

Activity 9.4: THINKING and CONCLUSIONS

Some answers are given, some you must determine, but all your answers should be justified!

1. Can you read the following message: ICURICUBICUR2YS4ME? (Hint: Read the letters!)

2. I am thinking of a counting number less than 100 and it is divisible by 3, by 4 and by 2 with a remainder of 1, but also by 5 with a remainder of 0. What is the number?

3. Geometry problems:
    a. What is the length of the hypotenuse of an isosceles right triangle with side s?

b. What is the diagonal of a square with a side of S? Answer: S√2

c. The diagonal of a cube with a side S is?

d. If we had a 4 dimensional cube, what would you predict the diagonal would be?

4. What is the sum of the first 15 even counting numbers or digits?

  a. 2
  b. 2 + 4 = 6
  c. 2 + 4 + 6 =12
  d. 2 + 4 + 6 + 8 = ?
  e. If you see the method you don't need to continue, just write the sum for the first even 15 digits.

5. Draw 6 circles and make each circle bigger than the preceding one.

a. On the first circle put 2 points, on the second circle, put 3 points, the third put 4 points, etc. and the last circle will then have 7 points. Now for each circle, connect the points and count the number of segments on the circle. As you do this, predict the number of segments on the next circle and repeat the process.

b. Can you state the formula for the number of segments given n points? Check your answers for cases n = 6, 7, and 8.
c. Now do the same for the number of regions.
d. **This exercise using regions will show you the weaknesses in unproven generalizations.**

6. An ordered movement problem:

s
M
L

———   ———   ———

1   2   3

The object is to move all 3 letters that are currently in column 1, to column 3, by x moves, one letter at a time, and never have a larger letter on top or above a smaller letter. The letters may move multiple times in different columns, just make sure that a larger letter is never on top or above a smaller letter.

7. Eight cell phones are identical except one is heavier. How can you tell the heavier one using a balance type scale once or twice?

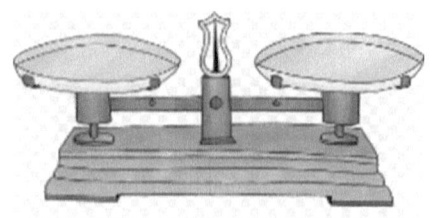

8. How did Galileo conclude that heavier and lighter objects fall at the same rate? The assumption at that time was that heavier objects fall faster. Hint: What was his thinking to justify they fall at the same rate? What are the 3 possibilities? Put the 2 objects together and how will the falling weights be altered?

9. Justify that the square root of 2 cannot be equal to an even integer or an odd integer. (Hint: The cases are the square root of 2 is equal to one of following: an integer, or is equal to an even integer.) Can you explain that each of the above cases lead to a contradiction?

10. The auto dealers said very few accidents are reported at speeds over 100ph. Conclusion is?

11. A true statement is: If you fly the American Flag on the 4th of July then you are patriotic. Write the Converse, Inverse and the contrapositive and classify each as true, false, or possibly true or false.

12. Check your newspaper or magazines for questionable ads and what they imply. Each person can find one ad and explain its implication.

13. Given: $x = a$ and $x^2 = xa$,
    then $x^2 - a^2 = xa - a^2$
    $(x-a)(x+a) = a(x-a)$

(x+a) = a
but x=a, therefore 2a = 1a
or 2 = 1.
Where is the error?

14. How would you design the field of stars
    in the U.S. Flag if we have 53 states?
    Hint: 8 and 7.

    (Good time to ask someone to read and
    report on the history of the flag.)

15. We use a base 10 counting system every
    day but the computer uses base 2.

Position Values

Base   10   **Base 2**
1000 100 10 1   32 16 8 4 2 1

Example:

Base   10   **Base 2**
Age   39   1 0 0 1 1 1
Age   61   ?
Write your age in base 2.

Family discussion: pertaining to your school
years. What do you recall from grades 1-6?

Grades 7-12? College? This can be very interesting
for the young members to hear from their parents,
plus their answers.

Activity 9.5: More Thinking

1. 12 hour and 24 hour clock times. Answer the following for` 12 hour time and 24 hour time:
    a. 3am + 5 hours = ? 3am + 5 hours = ?
    b. 7am + 2 hours = ? 7am + 2 hours = ?
    c. 8am + 8 hours = ? 8am + 8 hours = ?
    d. 10pm + 3 hours=? 10am + 15 hours = ?

2. If your work is 20 miles away, and you drive there in 30 minutes, what is your average miles per hour? (Answer is 40 miles/hour.)
    a. If your time to drive home from work is 40 minutes, what is the average miles per hour for the trip home? (Answer: 20 miles/40 minutes = x/60min and solving for x gives x = 30 mph.)
    b. What is the average speed for the round trip?

3. Two students home from college for the months of June, July, and August found jobs that paid the same amounts. At the end of June, student A was given a 5% increase, and Student B took a 10% decrease. Then, at the end of July student A took a 10% decrease and student B was given a 10% increase.
    Each student made $2000 in June, so what did each student make during the summer?

4. Draw a square with side s.
   a. What is the measure of the diagonal?
   b. Draw a cube with S. (Put in dotted line segments for hidden segments.)
   c. What is the measure of the diagonal of the cube?
   d. What is a tesseract?

5. Is 12345768990 divisible by 9?

6. All decisions are based on 4 items. What are they?

7. Draw a large circle.
   a. Indicate points A and B on the circle and draw chord AB. (2 points on the circumference of a circle and the connecting segment is called a chord.) Draw a circle with a chord.
   b. Add point C on the circle and draw chord AC, AB and BC. (3 points and three chords.)
   c. Add point D and draw and count the chords. The 4 points and 6 chords are: AB,AC,AD,BC,BD,CD)
   d. Add point E and draw the chords.
   e. Add point F and predict the number of chords and then count them.
   f. Do you see the pattern relating to points and the number of chords?
   g. Try doing this when points G and H are added and see if the pattern is correct. Did your indirect reasoning fail?

8. Taxi cab problem:
   a. One block 2 ways to drive

   B

   A

   b. 2 blocks and how many ways?

   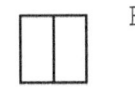
   B

   A

   c. Draw 3 blocks and how many ways?
   d. Draw 4 blocks and how many ways?
   e. Draw 5 blocks and how many ways
   f. Draw 6 blocks, your prediction is?

9. The diagonal problem:
   a. Draw a 4 sided polygon like a square.
   b. It has how many diagonals? Answer: 2
   c. Same type polygon as in **a** but with 5 sides, now how many diagonals? Answer: ?
   d. Same problem as b, but with 6 sides?
   e. Same as above, but with 7 sides. #?
   f. Same as above, but with 8 sides. #?
   g. Write your conclusion.

10. How many ways can pictures be hung horizontally?
    a. 2 pictures (Answer: 2 ways)
    b. 3 pictures
    c. 4 pictures
    d. 5 Pictures
    e. Predict for 6 picture?
    f. Prediction for **n** pictures

11. Gate problem. The following drawing shows a problem gate. What would you add to

it, to fix it? In other words, what would you add to correct the gate or make the gate into a rectangle? Did you actually use triangles to make this into a rigid figure?

12. Monetary Problem: You plan to prepare for your son's or daughter's college education. The plan is to invest $1000 per year for 10 years. Year 1: $1000 at 12% which is $1000(1.12)^1$ = **$1120.** Year 2 amounts to $1254.44. There must be an easier way instead of the above for all 10 years! There is. Do you recall the formula? **A + L method, and the formula is T = N(A + L)/2** (See Chapter 4, if you need a refresher.) L is the amount that the last payment will grow to after one year of interest, which will be 1120. A is the amount that the first payment will grow to after 10 years, which is 3105.85. (Use your calculator.) Now add those two together, A+L is $4225.85. The equation is 10{(A + L)/2}. The total is $21,129.25.

Family Discussion: Take some time to discuss finances and investments with your family. Topics can include banks, insurance, owning a house or renting, education, retirement and others. Each of these that you have should be discussed because they are very important! Invite one of your insurance or investment agents to come and speak to your group and answer any questions. Review some of your insurance policies yearly summaries as to value, performance, and contract terms, like whether you can terminate the policy without penalties. Some policies have a complicated set of terms, and many fees, such as a charge of as much as 20% of the check amount to write a check!

COMMENTS:

Chapter 10: Critical Thinking

Activity 10.1: Practical Thinking Problems

1. It is reported that most car accidents occur within 20 miles of a person's home. What are a few possible conclusions?

2. If the lights are out and you need a pair of socks, and you recall that in the drawer there is one pair of white socks and one pair of black socks, but not together or in pairs.

   a. What is the probability of selecting a matching pair in 2 selections? A selection is 1 sock.
   b. The probability of selecting a pair (WW or BB) in 2 selections?
   c. What is the probability of having a pair of white socks in 2 selections?

3. Given a pair of dice, what is the probability of throwing two 3s in 1 throw?

4. Using the information from problem 3, what should the fair bet payout be if a throw cost $2.

5. There are two types of probabilities, theoretical or empirical. The theoretical is from theory and the empirical from actual cases. In theory, the larger the number of cases, the probabilities of each should be the same. (You could adapt to problem 2.)

6. What is the meaning of **and** in most legal cases? Replace **and** with **or** and answer the question. Give examples!

7. Explain how implications (the 4 forms) may be used falsely in ads. (Give examples.)

8. Why does a 4-legged chair or table wobble? Why does a person sometimes use a cane?

9. How many license plates can be made if each license plate contains a combination of six total characters, and those characters can be any combination of letters and numbers. Examples: Two possible license plates are:

```
┌─────────────┐        ┌─────────────┐
│  LM2  N7S   │        │  ABC  123   │
└─────────────┘        └─────────────┘
```

Discussion: Are there any letters or numbers that may not be used? How many plates are possible or can be made or needed in your State?

10. Recall what the word **and** means and the word **or** means in many cases.

Give a few examples.

Some of the above may provide family discussions.

Activity 10.2: Problems

1. Probably most of you have played checkers or a similar game on a board. The question is how many **squares** are on an 8 by 8 checker board? (Hint #1: 64 is not the answer.) (Hint #2: The

answer for a 2 by 2 board is 5. Don't count the rectangles, only squares!)

2. Three identical boxes are almost identical in weight, 2 boxes are for the girls and the other box is for the boys. They are also wrapped identically, but the boy's box is a bit heavier. Using a scale like the following, how did the coach identify the boy's box? In one weighing? In 2 weighings? Explain your answers.

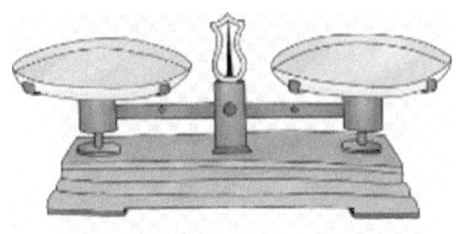

3. Galileo (History report: Who was he?) lived from 1564 – 1642 and justified that heavier objects and lighter objects fall at the same rate. The thinking at that time was that lighter objects fall slower than heavier objects. He then took 2 objects of different weights to the tower of Pisa, dropped them, and they did fall at the same rate. His conclusion was not verified from this, but by the following argument. He argued that if the 2 objects were tied together, then the new heavier object should fall faster (because it is heavier) or fall slower (because the lighter object should slow its fall), which is impossible, hence the only conclusion possible is they fall at the same rate. Consider the three cases and the only logical answer.

Family questions: Another interesting topic is to answer questions related to your teenage years and the same for the children who may now be in high school or college.

Activity 10.3: Critical Thinking

1. A teacher gave the following to the Superintendents, to justify that his afterschool activity should continue next semester.

The report:

   a. During the semester, the club membership increased 300%.
   b. During the semester we only lost 1 member.
   c. We are open to all students.
   d. If you were the superintendent, would you approve the club?

Answer: No, ask for more information, like the number of members in the club.

The reason is: What if the beginning membership was only 1, then what is the ending membership?

Many times, conclusions are suggested and approved without sufficient information!

2. A candidate running for office is asking for your vote. He or she will vote for lower taxes. What are a few questions you would like answered and explained before you decide to vote for this candidate?

3. Collect ads from magazines or newspapers and listen to the interpretations of the ads by your group.

Comments?

4. Three golfers after 18 holes were asked the results. Their answers were:
   a. Robert had the highest score.
   b. Ed did not have the highest score.
   c. Rich did not have the lowest score.
   d. (See page 29 for a possible method.)

They also said only one of the above is true. Hint: If you are not last, then you are either number 1 or 2.

Answer: Robert is #1
Ed was in third place.
Rich was in second place.

5. A weight problem, but a good buy.
The athletic department purchased 6 buckets of identical tennis balls. Each bucket was from a different company. One bucket contained defective balls and weighs less. It was a good buy so the coach decided to keep it. But they needed to know the defective bucket, so it will not be used in competition. How can the defective bucket of balls be identified in two weighings?

6. The following uses the concept of implications and the 4 forms, original A-B, implication, converse, inverse and contrapositive. Use the following diagram to understand.

A B (If in A then in B.)

A ➡ B (If in A then in B.)

Do you recall the interpretations for the converse, inverse and the contrapositive?

If so, then answer the following.

It was reported that very few accidents occur at speeds above 100mph. (Call this the A → B. Write this as a given statement.) Then write the 3 other forms, which many people assume to be true. (The forms are converse, inverse and the contrapositive).

Which ones may be false or invalid? (See diagram, above.)

7. It was reported that a 100 year old woman smoked her whole life. Conclusion: Therefore, smoking prolongs life. (S -> L) What are some fallacies with this inductive conclusion?

Family project: Create a history of your family, especially pictures. If some are in other states, contact them via email and request pictures and comments.

Chapter 11: Thinking and Conclusions

Activity 11.1: Review

1. a. It was reported that most car accidents occur within 20 miles of your home. Write this in if->then form and your possible conclusions. Write the 4 implications of this statement and your comments.

b. In a television ad, a pro golfer uses ball brand X. What conclusion does the ad hope you will assume? Write the 4 forms of the implication and comments.

c. One cigar ad always shows a man on a horse and smoking. What does the ad want you to assume? Write the 4 forms of the implication and your comments.

d. Ed told his father that John did not stand when the American Flag passed by in the parade, therefore he is not a good American. Write the 4 forms of the implication and the interpretations.

Activity 11.2: Thinking Review

1. What is deductive reasoning?
2. What is inductive reasoning?
3. Give an example of each.
4. Which is more valid?
5. Conclusions are based on 4 items. What are they and give examples?
6. JOHN'S LOCATION PROBLEM

John walks the same route every day to and from school and passes a market, a computer store, a park, and a theater but not in that order. Going to school he passes the park before the computer store, but after the post office. After school he passes the theater first and the market second.

Draw a line segment and locate his home and school at the end points. What is the arrangement of the businesses?

7. After Halloween weekend the information in the halls was that 4 students had possibly painted the windows in the Art Lab. The dean called the four suspected students in and they confessed that one of them did the painting. The Dean said let's play a little game and they agreed. The rule is that each of you will make a statement, but only one of the statements will be true. They talked a while and gave the dean the 4 statements. Here are the 4 statements. **(See page 29 for a method.)**

Seth: Chelsea did it.
Chelsea: Terri did it.
Eric: I did not do it.
Terri: Chelsea lied when she said I did it.

If only one of the statements is true, then who did it?

FAMILY HISTORY: Your city's history as to how it got the name. Another report could be the history of your local neighborhood, dates and some names of your hospitals, churches, schools and cemeteries, plus dates.

Chapter 12: Final Chapter after 12 months

Activity 12.1: Review of Thinking Material

1. A math problem, but very practical. Can you calculate the maximum number of license plates possible in your state, using only capital letters and numbers? Are some letters or numbers not used and why? (O and o is an example.) List any others. Assume each plate has spaces for 6 letters or numbers. How many possible plates can be made?

1A2B3C

2. Three students from the same school were in an accident, involving their car, but did not want to tell who was driving. The police asked each student to make a statement as to who was driving, but only one of the statements is to be true. (The police knew the students from the local high school and that they were taking Math.) He said he would try to solve it and if he couldn't they could help him. "Just wait a few minutes and I may have the answer as to who was driving the car and is responsible for the accident."
Here are the statements.
Max's: Bob did not drive the car.
Bob's: Joe did drive the car.
Joe's: Bob is lying.

Who was driving the car that damaged the pickup?

Answer: Bob drove the car.

3. A perfect number is one whose factors add to the number. The first perfect number is less than 10. What is it? The second is less than 30. What is it?

4. How many integers less than 100 have squares less than 100?

Family Activities: Have you ever had a family gathering for all your relatives, or at least invited the nearby relatives?

Project: Draw a family tree. (Add pictures from past years, if available.)

Activity 12.2: More Thinking Problems

1. Jaime is planning for a trip, but rain is predicted. If her suitcase measures 15 in. by 5 in.by 10 in., then what is the longest umbrella she can pack? (Hint: The answer is not 15.)
2. What do you recall about the relationship of the sides of a triangle? Given: Triangle measures 5, 6 and c. What do you know about the measure of side c or complete ? > c > ?
3. From the first 10 digits write a few possible measurement combinations for possible triangles. Example: A 3,4,5 right triangle has sides of 3 + 4 > 5

Write the measures (integers) for the possible lengths of the sides for each case. Example: Can you have a triangle with sides 2,7,6? or 2,7,8? or 2,7,10? Write your conclusions.

4. Some ad writers know what perfect numbers are and use their meanings in ads. A bottle of tablets was named NEW LIFE and you are to take 1 tablet PER DAY. How many tablets do you think are in the bottle? Two answers: 6 or 28. Why?

5. Another way to check answers. This method will work for the four operations.

Example:	1357 times 248 = 336536		
Add the digits:	16	14	26
Add again:	7 x	5	8
	7x5 =	35	8
	Sum	8	8

Try the method on an addition problem for better understanding of the method.

$$104 + 217 = 321$$
$$5 \qquad 10 \qquad 6$$

Adding to a one digit answer:
$$5 \quad 1 \quad = 6$$
Adding to one digits, 6 = 6

This tells you the final digital correct answer adds to 6. Try cases for subtraction and division.

6. Select a number between 50 and 90 but not a double digit number like 66. Add

the negative reverse of the number. (If the number is 62, then add a negative 26.) Keep repeating this method until you have one digit number. What is the number? Then, select another number from the given set and repeat the procedure until you have assumed a conclusion. Write your conclusion. Predict the result for a 3 digit number (like 538). Then, do it!

Activity 12.3: **The last set of problems**

1. Over the years, as cars and roads have improved, speeds have increased, but the understanding of braking distances have not. This topic can make for a discussion and better understanding, especially for people under 65. The case we will use is 70 mph speed. How far does a car travel in one second when doing 70 mph? Answer: 1.36 miles or 7072 ft. **Use a calculator and justify the answers!**(Why were people under 65 selected?)

2. Eric, Loren and Seth had a plan for some summer activities, and to fund these activities they agreed to make the following investments: Seth $1500, Loren $500 and $750 from Eric. At the end of the summer the profit was $2400. How should it be divided based on their investments and how much did each make (profit) during the summer?

3. Decisions or conclusions are based on or the result of:

    What you see? (TV)
    What you read?
    What you hear?
    National polls!
    Direct and Indirect reasoning!
    Traditions?
    Biased conclusions!

    a. What are the possible weaknesses of each of the above?
    b. What can you do to prevent false conclusions from each? (Example: One poll was based on only 814 adults and was rated as a national poll.)
    c. Your comments!

4. Many people listen to only a few TV channels or radio stations. Do you think these persons may have one sided or biased conclusions? What is your solution to insure valid decisions or conclusions?

5. A problem to show that conclusions can be False (Indirect Reasonings): Draw a 4 inch diameter circle on your paper with points A and B on the circle. Complete the following and you will be surprised! In the above case the result is 2 points connected and the circle is divided into 2 regions. Try to predict the number of regions before counting the number of regions.

	Points	Regions	
		Prediction	Actual
Case 3	3	?	?
Case 4	4	?	?
Case 5	5	?	?
Case 6	6	?	?
Case 7	7	?	?
Case 8	8	?	?

Conclusion: Based on a few cases, a general statement may appear true, but later with more cases tested it may show to be false.

6. Implications: Do you recall the four forms and their interpretations? Rate each of the following implications as to their Truth and Validity.

7. Given the statement A: If it is Thanksgiving Day, then we have turkey. True and valid?
   a. Converse: If we have turkey, then it is Thanksgiving Day. Not always true or valid.
   b. Write the inverse of A and classify it as possibly false.
   c. Do the same as in b for the contrapositive.
   d. Make a few If-Then statements from ads and discuss them.

8. Write the 3 forms of A->B, converse, inverse, contrapositive, for each of the following if-then statements and indicate their truth or falsity. Discuss your

conclusion. (A -> B is: If you have a drivers license, then you are over 21.)

   a. If you are over 21, then you have a driver's license.

   b. The following is the inverse of what implication: If you don't vote in your State elections, then you are not a citizen?

9. Conclusions and Laws are all based on four important terms, what are they?

10. Write the 4 forms of the following statement and indicate the truth and validity:

A->B. If you drive a car then you have a valid driver's license.

11. A thinking problem!

In my geometry class one day I tried to find which one of my top three students is the best thinker. Here is the problem. Each student was put in a corner of the room and was told a white or black hat will be put their head, while their eyes are closed. When they open their eyes, they are to put their hand up in the air if they see a black hat on either or both of the other two participants. They should keep their hand up, but if they can deduce the color of their own hat, they should lower their hand and state which colored hat is theirs. They better be correct about the color of their hat, because a free college education will be given to the student who correctly deduces the color of their hat.

A black hat was put on each one and the blind folds removed. In a few minutes one student lowered his hand and said "My hat is black!" How did he know?

Try this exercise with your friends or family. What do you learn from it?

Interesting Comment: In the late 1940s a professor at the U of Illinois named a new number. The name is GOOGOL. What is the number?

(The professor was a grandfather and was babysitting one day and asked the baby what is the name of this constant? The reply sounded like googol. True story? I don't know.)

Activity 12.4: The Last Word

It is the author's assumption that you and your family would conclude that this material would not only make your family members better thinkers, but also made your family one with better internal relations, for your children and for their children.

Plus, it may have been even enjoyable at times. Reminder: **Keep Asking Questions!**

Now that you have thought more about critical thinking skills as they apply to our everyday life, what areas of your life will you apply these skills? When reading news articles? Watching advertisements?

Should our elected representatives, State and Federal provide more news for critical thinking education as to how they voted. Reporting needs to be improved with explanations!

The author repeats his thanks to Jaime and Olaf for all their rational and clarifying help.

INTERESTING COMMENT: Each year, since 2015, a post graduate program has educated about 100 selected individuals in a program for persons who would like to be involved in Congressional programs for a better government. In the first class (about 100) 10% were West Point Grads, but about 1000 applied. The program was created by two former presidents to improve decisions by future governments. Presidents Clinton and the second Bush, plus the foundations of 4 former Presidents, Bush #1, Bush #2, Clinton, and LB Johnson to support the program (2 Republican and 2 Democrats). At one time, a graduation program for this group was held each summer in Washington D.C.

www.ingramcontent.com/pod-product-compliance
Lightning Source LLC
Chambersburg PA
CBHW021447210526
45463CB00002B/672